near FUTURE

Introduction to our changing world

Readers Knight

Copyright

2020 Readers Knight

All rights reserved.

Special Thanks

Ange Noir

&

Mathew Adenowo

This Book is a result of Enormous help

From these friends.

TABLE OF CONTENT

Introduction	7
Biology is our next best friend	18
Genes Are Magnificent	35
Impact On Life	48
Our Next Adventure	65
Finding The Planets	82
Conclusion	94

PREFACE

While the future can never be predicted with absolute certainty, present understanding in various scientific fields allows for the prediction of some far-future events, if only in the broadest outline. These fields include astrophysics, which has revealed how planets and stars form, interact, and die; particle physics, which has revealed how matter behaves at the smallest scales; evolutionary biology, which predicts how life will evolve over time; and plate tectonics, which shows how continents shift over millennia. This book will take on a journey to the future, which to be honest, is quite unpredictable. This book will go threw 6 Chapters.1. Introduction (To our Future)2. Biology is our next best friend (Genetics)3. Genes are Magnificent (CRISPR)4. Impact on Life (Applications of CRISPR)5. Our Next Adventure (Megastructures of Space)6. Finding the Planets (Best fit for our Accomodation)The Near Future is a series of 7 books. Each of 50-60 pages long, and this first part is the general introduction of our upcoming future. Other books will go through different chapters from this first book.

INTRODUCTION

To Our Future

We wake up every morning, do our chores, live our daily lives…always in the back of our minds wondering, what the future holds- and that is exactly what being a spacefaring civilization is all about.

It is about the belief & hope in all out hearts, that the future is going to be better. In so many ways. Better than the past. Better than the present. Better in all respects, be it education, medicine, science, justice, security, travel, marketing, food; we expect it to be *better*.

And I personally, cannot help but be ecstatic over the most amazing possibility of intergalactic travel!

I mean, just staring at the full moon on a clear night gives us so much peace, so much tranquility…how surreal could it be, if we were on it?! All of us, with our loved ones, having a picnic or a romantic dinner or a hangout with friends?!

That possibility...the vision of it, gives me goosebumps everytime.

And so, I ask we, dear reader:

What does *our* future hold? What *better* does our 'then' have, that our '**now**' lacks?

Here is a swish view, on how our lives are going to be, in the Near future.

A 'POST HUMAN' FUTURE?

While many researchers recognize that human biotechnology might pose some if not more serious demerits to the human population, from an individual to the entire society, others tend to downplay these challenges.

Some embrace the vision of a 'post-human' or 'transhumanist' future with all their heart, where people as well as other species could be so dramatically transformed that they are no longer the classics that we are accustomed to come across as of now.

While this is merely amarginal view, transhumanist advocates are vocal proponents of 'enhancing' physical and cognitive abilities through genetic modifications as well as various implants and other complex surgeries, in the sole hope of transcending the negative aspects of life, such as aging and dying.

From their perspective, the fact that many applications of biotechnologies for these purposes would likely exacerbate existing inequalities is typically of little concern if not at all.

PLANNING IT

Forecasting the future, at least for now, is impossible; making specific, assured predictions, absurd. However, to learn all that we can, and plan a significant change, we must look beyond the limits of our day to day discipline and short-term goals. We shall invariably consider a broad host of factors, and let our forecasting, influence our present.

Five trends shall 'shape the next 50 years':

First, much of the industrialized world's population is getting old, ageing, that is.

Second, political and economic power shall move from the West to the East, and military power would undoubtedly follow.

Third, global connections shall multiply several folds. More & more people will go online, and economies would become deeply interwoven.

Fourth, a quartet of technologies:

Genetics

Robotics

Internet and

Nanotechnology

(GRIN) – shall be able to create a new & interesting market of self-replicating machines (the kind that we can all recollect seeing in one of the biggest classics of all time, *The terminator*; I seriously loved that movie!), artificial intelligence and, my personal favourite, genetically modified species.

All these four, shall come together to interact with the fifth & final trend, the increasing importance of ecological concerns.

CHANGING LIVES

Though broad trends shall sweep across the globe, people couldn't still possibly lead uniform lives.

Instead, they would nhabit 'multiple futures'. Societies would polarize even farther, pulling people to opposite directions. Globalization would spread; ideas and intellectuals would flow freely in the atmosphere!

'Localization', the counter-trend to globalization, would signal a 'new tribalism' wherein people shall seek identity in their local communities, regions and nations.

The frenetic pace of the future could make a lot of people a bit anxious and overwhelmed. They could suffer from 'too much information' (TMI) or 'too much choice' (TMC) or both. More individuals would live alone due to a variety of options, longer life spans and lower birthrates, so they would indulge in more electronic entertainment and networking. Online, nothing would be 'neutral'.

Environmental concerns would affect transportation. Shrinking oil reserves and looming climate change would force the auto industry toward alternatives such as 'small electric and hybrid vehicles'.

Transportation definitely would get smarter: They could check our pulse, turn on using fngerprint and iris recognition, and even drive themselves.

What's more, automonitoring technologies would provide black-box data on accidents, occupants and driving routes. As the world gets more crowded, people would focus on public transport & carpooling, renting them as needed rather than owning them full time.

The political landscape would change as power shifts eastward, especially to China. Widening divides between the rich and the poor shall promote enclave living, and large urban areas might devolve to the 'feral' poor. Water shortages could push continents & nations to mutual wars, if an alternative isn't available till then.

Advances in genetic technology would help scientists predict criminal tendencies, and DNA samples stored in national databases would allow governments to track individuals' movements.

Combine these advances with the fear of terrorism and general anxiety about change, and the result is a lot more stress.

People will deal with it by retreating into polarized worldviews; they would elect politicians who cater to their local, specific issues, not necessarily the best candidates for the bigger picture. Short sighted visions would be more stressed upon by the laymen.

A nervous populace would handle more personal needs online, like shopping, learning and voting, rather than interact with others.

Some will opt to stay home for vacation, choosing to retreat into ever more elaborate virtual worlds. Others will travel more, and as the East continues to industrialize, its tourists will join those from the West. As a result, all the great historical sites and natural wonders would have waiting lists, and remote locations will become more popular as getaways. With the world increasingly linked, we'd be able to forward our choices of food, ambiance and lodging electronically, so our hotels would be ready with rooms and menus organized to our liking.

Dichotomies would also show up in the future of food. On one hand, life's increased pace would lead many people to prize speed and convenience. Food would increasingly arrive pre-packaged, pre-washed and pre-cut, and we'd be able to buy it faster due to electronic labeling. We would know more about our food, because tracking technologies would trace it from the farm to the plate. On the other hand, as part of society's pushback against technology, we'd seek locally grown food as well as comfort food that reminds us of simpler times. Some restaurants would move away from offering numerous choices to narrowing their selections: They'd choose for us.

Technological and scientific advancements would lead to genetically modified crops, as well as farmaceuticals and nutraceuticals: foods adapted to include extra nutrition, medications or biochemical additives tailored to our genetic makeup. These enhancements would mean that our diet would dictate ourr insurance and link to our computerized medical records. We could calculate our taxes based on the energy costs of our food.

Technology has the power to do many things, and changing the world is one of them.

We're privileged to be living in a time where science and technology can assist us, make our lives easier and rethink the ways we go about our daily lives.

The technology we're already exposed and accustomed to has paved the way for us to innovate further, and this list of current and future technologies certainly have the potential to change our lives even more.

SCIENCE & DEVELOPMENT

Society determines how far science can move forward. Specifically, biotechnology will produce new foods, creatures and tools. Genetics would allow us to manage our health better, even foretelling specific illnesses long before any physical symptoms appear.

The emotionally aware machines of science fiction would become a reality, as would a range of simpler robots. In general, machines will be more fully integrated into all aspects of life.

Computers would make increasingly accurate predictions in many areas.

Nanotechnology would cut across all industries, enabling synergistic change.

For example, minutely small sensors would be installed everywhere, allowing continual global monitoring and reducing the threat of accidents or even of getting lost.

Individuals would use these technologies to retrieve elusive memories. Groups would create knowledge and truth in new ways: Wikipedia is a harbinger. Ethical and political concerns will attempt to put a brake on scientific progress, but won't succeed until some technological disaster happens.

Science will advance over the next 50 years in response to future needs. For example, an aging population would fund more studies of elder care, biochemical memory aids and organ transplants.

Research into memory restoration, ostensibly to fight Alzheimer's disease, will develop transplants to erase recent painful memories. Hospital costs would continue to climb, thus spawning advances in telemedicine and distance monitoring technologies.

Scientific progress would also drive change in disturbing directions: Computers will become more intelligent than people by about 2030.

Imagine a world where parents can design babies who are intelligent, good-looking, and free from any hereditary disease.

These desires might be realized if the scientific advancements in genetic editing technology continue at an accelerated pace and our civilization resolves the ethical dilemma associated with the potential abuse of this technology.

CRISPR (Clustered Regularly Interspaced Short Palindromic Repeats) is a gene-editing technology that utilizes the Cas9 protein (it should be noted that since the making of this paper, a new, more efficient and precise protein called Cas12a has been discovered) ("CRISPR-Cas12a More Precise Than CRISPR-Cas9").

CRISPR works by using three main tools: A GPS for locating DNA, scissors for cutting DNA (deleting segments), and a pen for writing new DNA (inserting segments).

A repair template tailored to specific parts of a DNA sequence guides Cas9. CRISPR sequences are made into short RNA (ribonucleic acid) sequences in order to match the DNA sequences. After an action similar to a GPS, the DNA is found. Then, Cas9 connects to and cuts the DNA, which shuts down the gene. Researchers can test the genome's function by using modified versions of the protein Cas9 that do not run the risk of actually cutting through the DNA.

~*~

BIOLOGY IS OUR NEXT BEST FRIEND

GENETICS

For years we've been able to eradicate illnesses, but in the Near Future, it looks like gene selection is moving towards enhancement – potentially creating a generation of extremely athletic, super-intelligent humans

It has been recognized for approximately a century that genetic factors play a role in human disease, but until recently genetics was perceived as focusing only on rare disorders. Despite major advances such as chromosomal analysis, prenatal diagnosis, and newborn screening, genetics has played a minor role in day-to-day primary care. Recent advances, especially the sequencing of the human genome, are changing this picture rapidly. It is expected that genetics will play an increasingly central role in all areas of medicine and particularly in primary care, as the genetic contributions to common disorders come into focus.

Most practicing physicians have had little training in genetics, and rapid advances in the field make it increasingly difficult to keep up. This chapter reviews some of the basic principles of genetics to provide a foundation for understanding the application of genetics to medical practice.

GENES & GENOME

Our understanding of the concept of the gene evolved over most of the twentieth century, beginning with the recognition that genes function to encode proteins. The structure of DNA came to light in the middle of the century, ushering in a period during which the mechanisms of gene replication and expression came under study. The last quarter of the century saw the introduction of methods of DNA sequence analysis, culminating in the determination of the human genome sequence just at the turn of the century.

This section reviews the current understanding of the structure and function of genes and their organization into the genome.

STRUCTURE & FUNCTION

The basic unit of genetic function is the gene, the chemical basis for which is the DNA molecule.

DNA consists of a pair of strands of a sugar-phosphate backbone attached to a set of pyrimidine and purine bases.

The strands are held together by hydrogen bonds between adenine and thymine bases and between guanine and cytosine bases.

Together these strands form a double helix. The strands separate during DNA replication, and the base sequence of the newly synthesized strand is dictated by the complementarity of adenine with thymine and guanine with cytosine. DNA therefore contains within its structure the information necessary for its replication.

The sequence of bases in DNA also provides the code that determines the structure of proteins. Proteins consist of chains of amino acids. The specific ordering of amino acids determines the unique properties of each protein

THE HUMAN GENOME

The past decade has seen the completion of an international effort to sequence the human genome. Analysis of the sequence data has yielded several important insights.

The first is that there are fewer genes in the human genome than had been expected.

Early estimates of the number of human genes hovered at approximately 100,000; this was a crude estimate based on the facts that there are 3 billion base pairs of DNA and an average gene is approximately 30,000 bases.

GENE MUTATION

The properties of a protein are determined by its amino acid sequence, which in turn is determined by the base sequence of the gene that encodes it.

Alterations in the DNA coding sequence are referred to as mutations.

Mutations can lead to complete failure of expression of a gene, aberrant regulation, or abnormal function of protein products of the gene.

PATTERNS OF GENETIC TRANSMISSION

An individual inherits two copies of every gene (except those carried on the sex chromosomes), one copy from each parent. The copies of a particular gene are called alleles. The specific pair of alleles at a locus constitutes the genotype, and the resulting physical trait is the phenotype. Each human cell contains 46 chromosomes, including 22 pairs of non-sex chromosomes and two sex chromosomes (XX in females, XY in males).

SINGLE GENE TRANSMISSION

A recessive trait is expressed only in individuals who inherit a mutant allele from both parents.

Such an individual is said to be homozygous, whereas the parents are heterozygous carriers. Carriers do not express the trait because the non-mutant allele is dominant. Two carrier parents face a one in four chance that any of their offspring will be homozygous.

HOW GENES WORK

Living things inherit traits from their parents. People have known for many years that That commonsense observation led to agriculture, the purposeful breeding and cultivation of animals and plants for desirable characteristics. Firming up the details took quite some time, though. Researchers did not understand exactly how traits were passed to the next generation until the middle of the 20th century.

Now it is clear that genes are what carry our traits through generations and that genes are made of deoxyribonucleic acid (DNA). But genes themselves don't do the actual work.

Rather, they serve as instruction books for making functional molecules such as ribonucleic acid (RNA) and proteins, which perform the chemical reactions in our bodies.

Proteins do many other things, too. They provide the body's main building materials, forming the cell's architecture and structural components. But one thing proteins can't do is make copies of themselves. When a cell needs more proteins, it uses the manufacturing instructions coded in DNA.

The DNA code of a gene—the sequence of its individual DNA building blocks, labeled A (adenine), T (thymine), C (cytosine) and G (guanine) and collectively called nucleotides— spells out the exact order of a protein's building blocks, amino acids.

Occasionally, there is a kind of typographical error in a gene's DNA sequence. This mistake— which can be a change, gap or duplication—is called a mutation.

A mutation can cause a gene to encode a protein that works incorrectly or that doesn't work at all. Sometimes, the error means that no protein is made.

But not all DNA changes are harmful. Some mutations have no effect, and others produce new versions of proteins that may give a survival advantage to the organisms that have them.

Genetics is moving rapidly from a position at the fringes of medical practice toward the center.

Although the transformation will be gradual, physicians will be using the tools of genetics and genomics at virtually every phase of the medical decision-making process.

Some familiarity with the basic concepts of genetics is therefore essential to insure that these tools are used well and wisely.

There is also a need for access to point-of-care information and decision-making support.

Although genetics is likely to be integrated into medical education for the current generation of students, there remains a significant challenge in providing up-to-date information and understanding to physicians currently in practice who would be witness to major advances during the balance of their careers.

HOW SCIENCE HAS CHANGED OUR LIVES

If we look life 100 years ago, and compare that with the today's life, we will notice that Science has dramatically changed human life. With the dawn of the Industrial Revolution in the 18th century, the effect of Science on human life rapidly changed. Today, science has a profound effect on the way we live, largely through technology, the use of scientific knowledge for practical purposes.

NEAR FUTURE

Some forms of scientific inventions have changed our lives entirely. For example the refrigerator has played a major role in maintaining public health ever since its invention. The first automobile, dating from the 1880s, made use of many advances in physics, mathematics and engineering; the first electronic computers emerged in the 1940s from simultaneous advances in electronics, physics and mathematics. Today we have extra high- speed super computers with 100 % accuracy.

Science has enormous influence on our lives. It provides the basis of much of modern technology - the tools, materials, techniques, and sources of power that make our lives and work easier. The discoveries of scientists also help to shape our views about ourselves and our place in the universe.

1. Research in food technology has created new ways of preserving and flavoring what we eat.
   Research in industrial chemistry has created a vast range of plastics and other synthetic materials, which have thousands of uses in the home and in industry. Synthetic materials are easily formed into complex shapes and can be used to make machines, electrical, and automotive parts, scientific, technical and industrial instruments, decorative objects, containers, packing materials and many other items.

2. The use of science in daily life has helped us good deal in solving problems, dealing with the maintenance of health, production and preservation of food, construction of houses and providing communication and trans-portational (related to transport) facilities. With the help of Science we have controlled epidemics and much other kind of diseases. Now we know the basic structure of DNA and Genetic Engineering is conducting research to find out the right and correct Gene Therapy to overcome all the diseases.

3. Science has changed the people and their living life style, food habits, sleeping arrangements, earning methods, the way of communication between people and recreational activities. All kinds of music systems, computer games, electronic video games, DVDs, cinema entertainment and communication have been brought to our door with the help of Science. The life of man was very different from what it used to be 100 years back. Science has given ears to the deaf, eyes to the blind and limbs to the crippled. Science has adequately, energetically and productively advanced, changed, civilized, enhanced and progressed human life. Science has brought sophistication to human life.

4. In short science has changed, improved, enhanced, modified and refined human life in all ways.

5. Today with the help of Science we can explain what was strange and mysterious for the people of the past. The Science of Genetics opening new doors of understanding the human gene and cell.

6. Now human beings have become more critical and less fearful than our fore-fathers and ancestors.

7. Two hundred years ago death rate among children was very high. In those days seven out of eight babies died before their first birthday. Now with the help of vaccines, medications and proper health care system life expectancy has improved. Now people live longer and safe lives as compared to 200 years ago. Biochemical research is responsible for the antibiotics and vaccinations that protect us from infectious diseases, and for a wide range of other drugs used to defeat specific health problems. As a result, the majority of people on the planet now live longer and healthier lives than ever before.

8. After that and up to the age of 12 one used to fall in a prey to diseases like small pox, measles, whooping- cough, scarlet fever and diphtheria. Now Science has defeated these diseases.

9. At a later stage again one was under constant threat of yellow fever, malaria, typhus, cholera, typhoid

and influenza. Today we have vaccines and medical aid to cope with these health problems. Further research is underway to find out the causes and treatment of these and other diseases.

10. From one person the disease used to spread among the other people. It is called Epidemics. Now with the help of Vaccines and Medications we have defeated these diseases. But still Science has to do more research and has to fight with other arenas of diseases.

11. Life was uncertain. It was rare to see to somebody thirty years old because due to diseases many people died earlier than the age of thirty. These conditions were prevailing just a short while ago.

12. In everyday life, we have to communicate with different friends and relatives, various official people and for general purposes. And many people to be contacted can be at very far off distances. However, time and distance both have been conquered by Science. Whether we want to communicate or travel, both are possible quickly, briskly and expeditiously.

13. These days there are very little chances of babies catching diseases, because births normally take place in hospitals under the supervision of a team

of specialist doctors. Science has invented vaccines for weng babies to protect them against future life illnesses.

14. Young people are also given medical treatment in time and these days the man lives for about seventy years.

15. Science and scientific methods have helped in finding out the cause of disease and its prevention.

16. Sanitary condition in the past was deplorable. Now we have better sanitary systems.

17. The city streets were unpaved; there was no proper drainage system. Garbage and other refuse was seen everywhere. Pigs were seen wandering through the streets. People got water from filthy wells. Now filtered mineral water is available to overcome diseases. Solid waste management is not a problem now a days, it is the duty of the city municipal committees to manage and dump it with the latest machinery and equipments

18. Now all these defects have gone. There is cleanliness everywhere. It is illegal to throw garbage into the streets. There is a proper drainage system and new and improved methods for solid waste man-

agement as it has been told earlier. There are separate departments that bother about sanitary condition of the towns.

19. A century ago for house hold purposes water was carried from wells outside in buckets. It sometimes proved injurious to human health. Moreover, it was insufficient for the daily needs. But now water filters have become a thing of common usage.

20. Now there is sufficient supply of water in cities. For example Los Angeles gets water through pipes from Colorado River, which is 340 miles away. This water is supplied to Los Angeles after the proper water filtration process.

21. With the help of science there is change in our food also. We get varieties of food. In the past, food could not be preserved. But now the quick freezing methods have made possible preservation possible. Due to modern technologies like dehydration and sterilization there is no chance of food poisoning. We get all kinds of fruits, meats and vegetables. Even those fruits and vegetables which are out of season.

22. Not only our eating habits are changed, but also there are improvements in our houses. Means of

transport has also undergone a big improvement and change.

23. Science has also changed our attitudes. Superstitions have been discarded, because there is no scientific basis for them. Now people do not fear cloud thunders.

24. Now people no more believe that diseases are caused by evil spirits.

25. Astrology and fortune- telling have lost popularity as compared to 100 years ago. Nobody now fears black cats, broken mirrors and the number 13. Because science has proved that these kinds of fears are un-scientific and illogical.

26. Science has changed the longstanding false notions of the people, which are not supported by Scientific Facts.

27. Research in the field of science and technology has made people open-minded and cosmopolitan, because the Scientist does not like to travel on the beaten track and he always tries to find out new things, new explorations, new discoveries and new inventions.

28. Science has also brought medical equipments that help to save human life. The kidney dialysis machine facilitates many people to survive kidney diseases that would once have proved fatal, and artificial valves allow sufferers of coronary heart disease to return to active living. Since the 1980s lasers have been used in the treatment of painful kidney stones. Lasers are used when kidney stones fail to pass through the body after several days, it provides a quick and low-pain way to break up the stone and allow the stones to be easily passed through the body. This technique is called Lithotripsy.

29. Arthroscopic surgery is a technique using fibre optics to probe complex joints such as knee, shoulder, ankle and wrist to evaluate injury. It is a minimally invasive operation to repair a damaged joint; the surgeon examines the joint with an "arthroscopy" while making repairs through a small incision.

30. 200 years ago nobody even knows that human body parts can be replaced or transplanted. Now kidney transplant is widely used to save human lives around the globe. Dr. Christian Bernard first of all invented the method of heart transplant. Eye transplant techniques are used in these days to see again this beautiful world, for those who have lost their eyes. These all are the blessings of Science.

31. Ultra-high-frequency (UHF) waves are allocated for variety of uses, including television, cellular phones, public safety radios, business radios, military aircraft communications, military radar, cordless phones, baby monitors, etc. So, whether someone is watching over-the-air TV, talking on cell phone, having police/fire/ambulance dispatched to an emergency they are experiencing, or having national airspace protected by military aircraft, they all are benefitting from the science that has allowed the use of UHF waves. Even it is used to treat some illnesses.

32. For communication, now we have fixed wire telephones, moveable wireless phone sets, cordless phones, mobile phones, wireless, video conferencing, Internet, Broad Band Internet, E-mail, Social Networks, Satellite Communication and many other ways to communicate. These all are blessings of Science. Today we are better aware of what is happening around the globe due to satellite television channels. The benign and benefits of science for human life are endless.

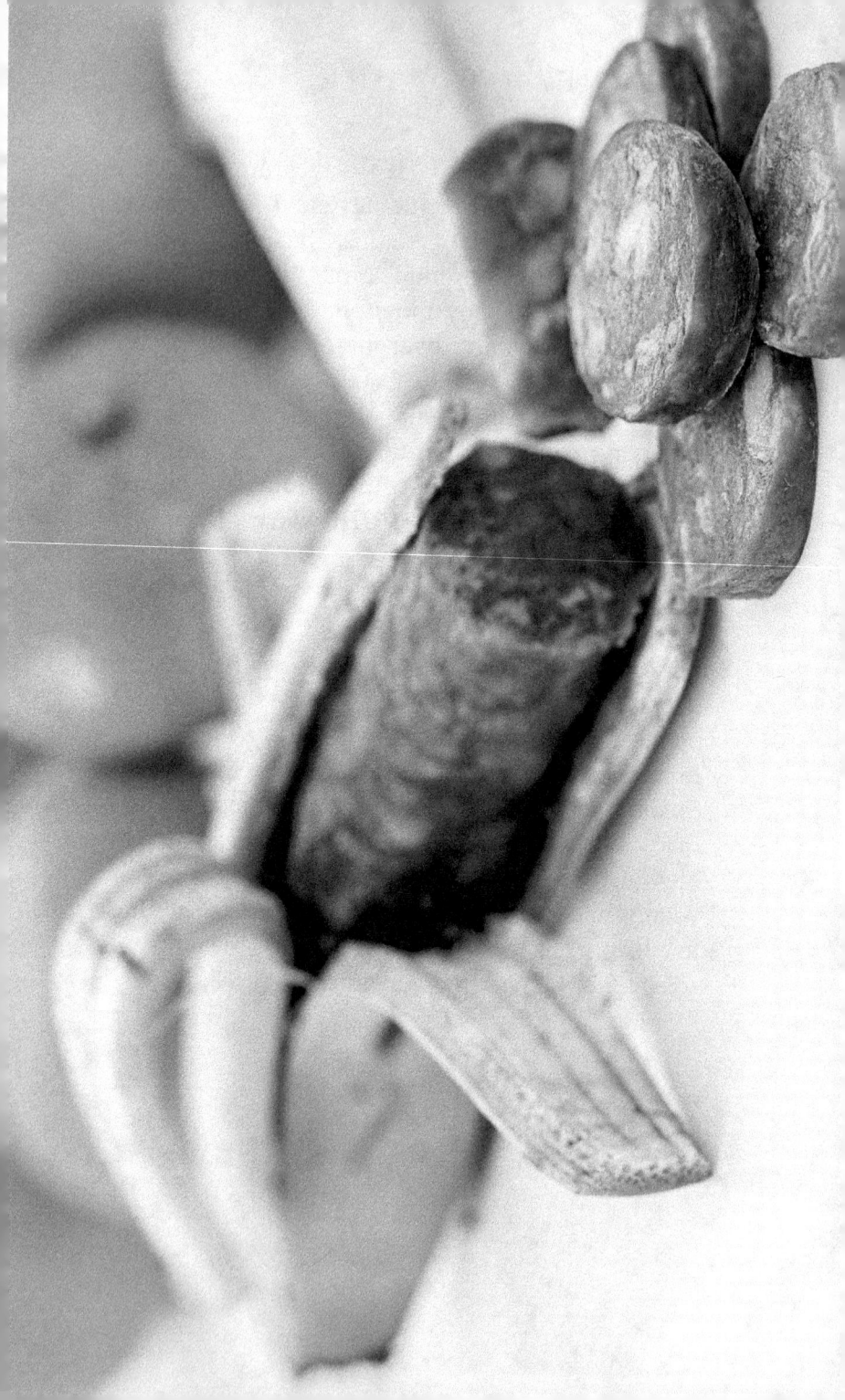

GENES ARE MAGNIFI-CENT

CRISPR

Designed human beings are just another out-of-science-fiction concept that we are getting closer to making it a reality.

The concept of designer babes has been discussed a lot in recent months after a Chinese doctor claimed he helped create two babies with modified genes. This has sparked various debates on the ethics of genetic manipulation and the future of genetics.

The term 'designer baby' refers to a baby that has been given special traits through genetic engineering. This is done by altering the genes of the egg, sperm, or the embryo. These traits can, in theory, vary from lower resistance to diseases to even gender selection.

WHAT IS GENE-EDITING?

Genetic editing is the process of making changes to the genetic code (DNA). In the case of 'designer babies,' this is done either by removing small sections of the existing genome or by introducing new segments of DNA into the genome.

A new technique, called CRISPR (clustered regularly interspaced short palindromic repeats) has allowed scientists to cheaply and very rapidly alter the genome of almost any organism. In the most common form of CRISPR, an enzyme called Cas-9 is used to cut out selected sections of DNA or add new sections to the existing DNA.

GENE-EDITING IN HUMANS: WHAT IS CONTROVERSIAL AND WHAT IS NOT?

Genetic editing in humans is a controversial topic, but not all forms of human genetic manipulation are in question. For example, CRISPR could be used to alter cells in the bodies' immune system in order to target and destroy cancer cells or to replace the genes that cause sickle cell anemia with non-sickle cell genes.

CRISPR is a tool with immense potential to create better crops and livestock, manufacture new drugs, eliminate pests, and treat critical illnesses. But the problem arises when there are no limits.

Gene-editing can be performed on both somatic cells and germ (stem) cells, and both these cell types offer very different results.

Somatic cells are those cells that have already differentiated into a specific type of cell, like a liver cell or a lung cell. Changes to these cells only affect the part of the body the cell belongs to, such as the liver or lungs. Any changes to somatic cells cannot be passed on to any offspring.

For this reason, the altering of somatic cells for the treatment of diseases is not generally regarded as controversial.

The problem arises when gene-editing is performed on germ cells. These are the cells of the egg or sperm, which eventually give rise to all the cells in the body. These cells can develop into any type of cell. This means that changes to the germ cells affect not only the child to be, but can also be passed on to future generations.

Germline cell editing is sometimes also referred to as embryo manipulation. Implanting a CRISPR-modified embryo into a human is illegal in some places and even where it is not illegal, it is in contravention of research guidelines, as well as moral and ethical standards.

THE ETHICS OF GENE-EDITING

The ethics of gene-editing can be viewed from a variety of different angles.

For many, the notion of experimenting on human beings is unethical, especially when there is insufficient evidence to suggest the experiment will be successful, or will not cause harm. At this point, there is not enough evidence to demonstrate that CRISPR is safe - we do not know the effects of editing any given gene on the existing person or on future generations. In fact, a recent study by the Wellcome Sanger Institute demonstrated that the use of CRISPR can lead to extensive genetic damage in the target genome.

For others, these techniques demonstrate a disregard of globally accepted scientific and ethical standards. These standards exist to prevent research on humans when there is a lack of evidence that it will be safe.

However, what worries many people most is the idea that in the future, parents or doctors will be able to dictate traits such as the gender, height, or intelligence of their baby - giving those who can afford gene-editing an advantage and potentially leading to a type of genetic class system. In essence, it would allow science and not nature to guide the evolution of the human race.

HOW DOES A DESIGNER BABY WORK

A designer baby is a baby whose genetic makeup has been selected or altered, often to include a particular gene or to remove genes associated with a disease. ... Other potential methods by which a baby's genetic information can be altered involve directly editing the genome – a person's genetic code – before birth.

DESIGNING OUR BABY

Imagine if you could choose what our brother or sister would look like. Imagine if we could choose our child's eye colour, hair colour, IQ level and even his interests. Imagine a world where everyone is perfect health-wise and IQ-wise.

This all seems like a scene from a science fiction movie right? Well, all this might be possible in the near future with the advancement of Pre-implantation Genetic Diagnosis or PGD. PGD is a procedure where the embryo is genetically tested to be free of disease and then implanted in the mother using invitro fertilization (IVF).

Parents can now choose to have the ultimate child, one who excels in both studies and sports. We all got a glimpse of PGD in the 1997 movie Gattaca where genetic engineering and IVF allowed for the engineering of children, including factors such as gender, intelligence, life-span and even eliminating various hereditary and genetic diseases.

However, PGD does not only exist in science fiction movies. In fact, PGD is continuing to develop into an exciting technology with the incredible potential to increase women's chances of not just a healthy pregnancy but a healthy child as well.

PGD was first made famous by Dr. Jeffrey Nisker, from the University of Ontario. He pioneered PGD and was the first man in Canada to offer such a revolutionary procedure of manually engineering embryos. He has since stopped and closed his programme down due to overwhelming requests from all over Canada to access PGD, not to avoid severe genetic conditions but for gender selection, primarily to avoid having a daughter.

It was against his concern and belief that all children should be cherished, regardless of their gender. Among the cases in which PGD was performed, was when a couple needed the procedure to have a baby with the right genetic make-up to save their son's life. Leanne and Stephen needed the procedure to save their first-born, BJ, who suffers from a rare genetic disorder, Hyper IGM. BJ is missing a vital part of his immune system and survives only with the aid of regular blood transfusions.

The child needs to be free of the disorder and of the same sex as BJ, then only can they be assured that the tissue can be transplanted from the newborn to BJ. The procedure was long and tedious with Leanne and Stephen going through a roller coaster of emotions as the eggs are extracted and fertilized and then sent for testing.

After a few rounds of misses, Leanne and Stephen made a hit and the embryo was transplanted in Leanne. Soon, their designer baby would arrive, bringing help and hope to his older brother and parents! PGD has, however, raised some ethical concerns from society, particularly regarding its potential and who should decide how to use this new found revolutionary technology.

The procedure allows for parents to determine the gender of the embryo, and can thus be used to select the embryo of one gender in preference to the other in the context of 'family balancing'.

HOW CRISPR WOULD HELP US WITH INTER-GALACTIC TRAVEL

Space is cold (mostly) and brutal, inhospitable to life as we know it. All other planets we know a lot about differ drastically from Earth, lacking the atmosphere we breathe and the gravity that we're accustomed to.

Yet as scientists and inventors from Steven Hawking to Elon Musk have pointed out, we may need to leave Earth and journey to new planets near and far if humanity is going to survive in the long term. We may need to find places to sustain our population and carry on in case we run out of resources here or in case the destructive power of our current technology leads to a disaster that makes life on Earth impossible.

Doing so would require difficult technological achievements, including developments in energy production, starship construction, telescopes, navigation equipment, and more.

This week NASA successfully tested the rocket motor that will one day propel astronauts beyond the relative safety of Earth and into the orbit of Mars.

The plan is to send astronauts to orbit Mars by the mid-2030s with a manned landing on the red planet following soon after.

The mystery and excitement of space has been engrained in society for decades now but the so-called 'Mars generation' brings its own challenges. Humans are very well adapted to life on Earth thanks to millions of years of living here, but the minute we leave it becomes clear how poorly adapted we, and the things that sustain us, are for life elsewhere.

Could CRISPR hold the key to some of the biggest practical challenges of human space exploration? Read on to explore how CRISPR could solve the problem of growing food off-planet.

USE OF CRISPR IN PLANTS

Excitement about CRISPR/Cas9 has largely focused on implications for human health: from the prospect of preventing genetic diseases to engineering the human immune system to fight off cancer without the use of medication. But can CRISPR/Cas9 be used in plants for genome engineering? We're a planet growing in population and receding in resource so if CRISPR can allow us to grow crops in places we currently think of as uninhabitable, this could solve problems not only here on Earth but in space too.

Use of CRISPR in plants is relatively recent: three studies published in the same issue of Nature Biotechnology in August 2013 were the first to conclusively demonstrate that Cas9 can be effectively used in plants cells.

- Scientists from the Chinese Academy of Sciences in Beijing disrupted endogenous genes in rice and wheat protoplasts as well as in rice calli

- Researchers at Harvard reported cleavage activity of Cas9 in protoplasts of Arabidopsis thaliana (a flowering plant native to Eurasia, the first plant to have its genome sequenced), and Nicotiana benthamiana (a close relative of tobacco indigenous to Australia)

- The Sainsbury Lab in the UK disrupted the phytoene desaturase (PDS) gene using Cas9 in Nicotiana benthamiana leaf tissue

- Those space plants look pretty new! They must be a few years away yet?

- The studies above formed the basis of numerous CRISPR/Cas9 edits in plants, proving that we can edit the genomes of plants in the same way as we have done with mammalian cells in the lab. More recent CRISPR/Cas9 studies in plants have demonstrated genome engineering of species that are practical crops for food production. For example, scientists in China reported the creation of a strain of wheat that is resistant to powdery mildew, a fungal disease that can affect and ultimately destroy a wide range of plant species.

There are also biotechnology-industry collaborations happening that aim to grow corn and wheat strains genetically engineered to resist drought. Such collaborations could transform the way we think about growing crops on Mars for example, a planet where drought is a constant. It's likely that we won't see drought resistant crops for a few years yet, but with NASA's aim of a Mars landing by the mid-2030s, that could fit in pretty well. Field trials are planned to begin in 2016 – we'll report back when preliminary results are released.

So plans to grow crops off-planet are in the works. But could we keep the food we grow fresh?

If we've managed to get over the first hurdle of understanding how to grow plants in such difficult conditions, it's critical that we keep our food sources fresh for as long as possible. After all, sustaining life away from Earth won't involve a trip to the supermarket when food stocks run low.

Researchers at Penn State University used CRISPR/Cas9 technology to create small deletions in a specific gene encoding a polyphenol oxidase (PPO). This enzyme catalyzes the oxidation of polyphenols in the mushroom when they are exposed to air, the first step in the production of dark pigments and what we know are 'browning'. Deletions in the gene for PPO gave a reduction in overall activity by more than 30%, meaning the 'shrooms stay fresher for longer and don't need to be handled quite as delicately as their un-edited counterparts.

Not so keen on mushrooms? Genetic engineering has been shown to prevent the browning process in both apples and potatoes too.

The approvals process for my "dinner of the future" is going to be a nightmare isn't it?

It doesn't look that way, at least not so far. These mushrooms have evaded the process of testing and approval by the US Department of Agriculture on the grounds that they don't contain any introduced genetic material, and therefore don't pose a threat to consumers or the environment in which they grow. Dr. Yang, who led the research team at Penn State, advised that the mushrooms won't be on shelves any time soon though because they may still be hit with regulations by the US Environmental Protection Agency or the Food and Drug Administration. We'll be keeping a close eye on this topic so keep an eye on the blog for more information as it happens.

Does all of this mean we'll one day be able to grow plants on Mars?

At the 2015 SynBioBeta Conference in San Francisco, NASA astronaut Catherine Coleman told the audience that when we go to Mars, we would need to have used Synthetic Biology to figure out how to grow our plants there for food. While there are other important factors to consider before we can do this – such as sources of water, fertilizer etc. – CRISPR/Cas9 engineering plants to help them adapt to harsh environments like those on Mars are bringing us closer to that reality.

IMPACT ON LIFE

APPLICATIONS OF CRISPR

CRISPR, the breakthrough method for editing genes, has the potential to improve our lives. But one of its inventors warns us scientists may be tempted to change life itself — in ways we won't like.

Genome editing is the deletion, insertion, replacement or modification of genes in a strand of DNA. It can be carried out by using a variety of DNA targeting tools such as zinc-finger nucleases (ZFNs), transcription activator-like effector nucleases (TALENs), or more recently, clustered regularly interspaced short palindromic repeats (CRISPR). Over the past 50 years, scientists have gained the ability to read, copy, synthesise and code DNA, and with new genome editing techniques, it is now possible to edit DNA in a precise and targeted fashion.

IMPACT TO OUR WELLNESS - 7 DISEASES CRISPR TECHNOLOGY COULD CURE

CRISPR technology offers the promise to cure any human genetic disease. Which are the candidates to be the first one?

CRISPR-Cas9 was first used as a gene editing tool in 2012. In just a few years, the technology has exploded in popularity thanks to its promise of making gene editing much faster, cheaper and easier than ever before.

CRISPR has already changed the way scientists do research. But what everyone is expecting, either with excitement or fear, is its use in humans. In theory, CRISPR technology could let us edit any genetic mutation at will, curing the disease it causes. In practice, we are just at the beginning of the development of CRISPR as a therapy and there are still many unknowns.

But if you had at least a chance to cure any genetic disease which one would it be? These are seven diseases that scientists are already tackling with the help of CRISPR Cas9, and which could eventually become the first conditions to ever be treated with this revolutionary technology.

1. Cancer

The first applications of CRISPR could be in cancer. Indeed, one of the first and most adva

nced CRISPR clinical trials, which is currently running in China, is testing the potential of the gene editing tool to treat patients with advanced cancer of the esophagus.

The treatment being tested at the Hangzhou Cancer Hospital starts with the extraction of immune T cells from the patient. Using CRISPR, the cells are modified to remove the gene that encodes for a protein called PD-1 — some tumors are able to bind to this protein on the surface of immune cell and instruct them not to attack. The modified cells are then reinfused into the patient with a higher capacity to attack cancer cells.

2. Blood disorders

The first CRISPR trial in Europe and the US, which enrolled its first patient in February this year, aims to treat beta-thalassemia and sickle cell disease, two blood disorders that affect oxygen transport in the blood. The therapy, developed by CRISPR Therapeutics and Vertex Pharmaceuticals, consists in harvesting bone marrow stem cells from the patient and using CRISPR technology to make them produce fetal hemoglobin, a natural form of the oxygen-carrying protein that binds oxygen much better than the adult form.

Before the trial started, the FDA put it on hold in the US to clear out some safety questions. A few months later, the hold was lifted and the treatment was given fast track designation for both conditions.

Hemophilia is another blood disorder that CRISPR technology could tackle. CRISPR Therapeutics is working with Casebia on an in vivo CRISPR therapy where the gene editing tool is delivered directly to the liver.

3. Blindness

CRISPR is a great candidate to treat genetic blindness. Many hereditary forms of blindness are caused by a specific mutation, making it easy to instruct CRISPR-Cas9 to target and modify a single gene.

In addition, the eye is an immunoprivileged part of the body, meaning that the immune system's activity is limited there. This becomes an advantage in sight of the concerns regarding the possibility that CRISPR could induce immune reactions against it, which would block its activity and derive into side effects.

Editas Medicine is working on a CRISPR therapy for Leber congenital amaurosis, the most common cause of inherited childhood blindness, for which there is no treatment. The company aims to target the most frequent mutation behind the disease, using CRISPR to restore the function of light-sensitive cells before the children lose sight completely.

4. AIDS

There are several ways CRISPR technology could help us in the fight against AIDS. One is using CRISPR to cut the DNA of the HIV virus out of its hiding place in the DNA of immune cells. This approach could be used to attack the virus in its hidden, inactive form, which is what makes it impossible for most therapies to completely get rid of the virus.

Another approach could make us resistant to HIV infections. Certain individuals are born with a natural resistance to HIV thanks to a mutation in a gene known as CCR5, which encodes for a protein on the surface of immune cells that HIV uses as an entry point to infect the cells. The mutation changes the structure of the protein so that the virus is no longer able to bind to it.

This approach was used in a very controversial case in China last year. CRISPR-Cas9 was used to edit human embryos to make them resistant to HIV infections. The experiment caused outrage among the scientific community, with some studies pointing out that the 'CRISPR babies' might be at a higher risk of dying younger. The general consensus seems to be that more research is needed before this approach can be used in humans.

5. Cystic fibrosis

Cystic fibrosis is a genetic disease that causes severe respiratory problems. Although there are treatments available to deal with the symptoms, the life expectancy for a person with this disease is only around 40 years. CRISPR technology could help us get to the origin of the problem by editing the mutations that cause cystic fibrosis, which are located in a gene called CFTR.

Researchers have proven that it is possible to use CRISPR in human lung cells derived from patients with cystic fibrosis and fix the most common mutation behind the disease. The next step would be testing it in humans, which both Editas Medicine and CRISPR Therapeutics plan to do.

However, cystic fibrosis can be caused by multiple different mutations in the CFTR gene, meaning that different CRISPR therapies will have to be developed for different genetic defects. Editas has stated that it will be looking at the most common mutations, as well as some of the rare ones for which there is no treatment.

6. Muscular dystrophy

Duchenne's muscular dystrophy is caused by mutations in the DMD gene, which encodes for a protein necessary for the contraction of muscles. Children born with this disease suffer progressive muscle degeneration, and there is currently no treatment available beyond palliative care.

Research in mice has shown CRISPR technology could be used to fix the multiple genetic mutations behind the disease. Last year, a group of researchers in the US revealed an innovative method that, instead of fixing each mutation individually, used CRISPR to cut at 12 strategic 'mutation hotspots' covering the majority of the estimated

3,000 different mutations that cause this muscular disease. A company called Exonics Therapeutics was spun out to further develop this approach.

Editas Medicine is also working in a CRISPR therapy for Duchenne's muscular dystrophy. It is also following a broader approach where instead of fixing mutations, CRISPR removes whole sections of the mutated protein, which makes the protein shorter but still functional.

7. Huntington's disease

Huntington's disease is a neurodegenerative condition with a strong genetic component. The disease is caused by an abnormal repetition of a certain DNA sequence within the huntingtin gene. The higher the number of copies, the earlier the disease will manifest itself.

Treating Huntington's could be tricky, as any off-target effects of CRISPR in the brain could have very dangerous consequences. To reduce the risk, scientists are looking at ways to tweak the gene editing tool to make it safer.

It's difficult to predict the outcome of these early efforts to use CRISPR as a therapy, but as these first attempts progress, more and more indications will certainly be added to the list. One of the biggest challenges to turn this research into real cures is the many unknowns regarding the potential risks of CRISPR therapy. Some scientists are concerned about possible off-target effects, immune reactions to the gene editing tool, or the fact that it could increase the risk of cancer. But only time will tell whether these challenges can be surmounted or not.

HOW CRISP CAN HELP US TO DESIGNER BABIES

This is the one that gets the most attention, and rightly so. It's not entirely far-fetched to think we might one day be confident enough in CRISPR's safe to use it to edit the human genome — to eliminate disease, or to select for athleticism or superior intelligence.

But Jiankui's recent attempt to introduce protection from HIV in embryos intended for pregnancy, which involved little oversight, is not how most scientists want the field to move forward.

The concern is that there are not enough safeguards yet to prevent harm nor enough knowledge to do definite good. In Jiankui's case, he also did not tell his university about his experiment ahead of time, likely did not fully inform the parents of the modified babies of the risks involved, and may have had a financial incentive from his two affiliated biotech companies.

We're not even close to the point where scientists could safely make the complex changes needed to, for instance, improve intelligence, in part because it involves so many genes. So don't go dreaming of Gattaca just yet.

"I think the reality is we don't understand enough yet about the human genome, how genes interact, which genes give rise to certain traits, in most cases, to enable editing for enhancement today," Doudna said in 2015. Still, she added: "That'll change over time."

BUT, TO BE, OR NOT TO BE?

Given all the fraught issues associated with gene editing, many scientists are advocating a slow approach here. They are also trying to keep the conversation about this technology open and transparent, build public trust, and avoid some of the mistakes that were made with GMOs. But with CRISPR's ease of use and low costs, it's challenging to keep rogue experiments in check.

In February 2017, a report from the National Academy of Sciences said that clinical trials could be greenlit in the future "for serious conditions under stringent oversight." But it also made clear that "genome editing for enhancement should not be allowed at this time."

Society still needs to grapple with all the ethical considerations at play here. For example, if we edited a germline, future generations wouldn't be able to opt out. Genetic changes might be difficult to undo. Even this stance has worried some researchers, like Francis Collins of the National Institutes of Health, who has said the US government would not fund any genomic editing of human embryos

A CRISPR FUTURE WAYS GENE EDITING WOULD TRANSFORM EARTH PLANET

Over the past few years, CRISPR has been making headlines. Experts predict that this gene editing technology will transform our planet, revolutionizing the societies we live in and the organisms we live alongside. Compared to other tools used for genetic engineering, CRISPR (also known by its more technical name, CRISPR-Cas9) is precise, cheap, easy to use, and remarkably powerful.

Scientists are still in the earliest stages of figuring out how we can use CRISPR to change the world for the better. Of course, the power to alter DNA — the source code of life itself — brings with it many ethical questions and concerns. With this in mind, here are some of the most exciting uses for this revolutionary technique, and the hurdles that might slow or prevent these technologies from reaching their full potential.

1. CRISPR Could Correct The Genetic Errors That Cause Disease

Hypertrophic cardiomyopathy (HCM) is a heart condition that affects roughly 1 in every 500 people worldwide. Its symptoms are painful and often deadly. Mutations in a number of dominant genes cause the heart tissues to stiffen, which can lead to chest pain, weakness, and, in severe cases, sudden cardiac arrest.

Off-target genetic mutations and mosaics (only some cells adopt the changes, meaning that a fraction of people would inherit the mutation) were only present in 13 of the 54 embryos.

To further reduce the chance that only some cells would be changed, the researchers carried out another experiment in which they corrected the same gene in embryos directly at the time of fertilization.

"By using this technique, it's possible to reduce the burden of this heritable disease on the family and eventually the human population.

2. CRISPR Can Eliminate The Microbes That Cause Disease

Though treatments for HIV have turned the infection from a virulent killer to a livable health condition, scientists still haven't found a cure.

That could change with CRISPR. In 2017, a team of Chinese researchers successfully increased resistance to HIV in mice by replicating a mutation of a gene that effectively prevents the virus from entering cells. For now, scientists are only conducting these experiments in animals, but there's reason to think the same methods could work in humans. The mutation that encourages HIV resistance naturally occurs in a small percentage of people. By using CRISPR to introduce the mutation to human stem cells that lack it, researchers could substantially bolster HIV resistance in humans in the future.

3. CRISPR Could Create New, Healthier Foods

CRISPR gene editing has proven to be promising in the field of agricultural research. Scientists from Cold Spring Harbor Laboratory in New York used the tool to increase the yield of tomato plants. The lab developed a method to edit the genes that determine tomato size, branching architecture and, ultimately, shape of the plant for a greater harvest.

High-yield crops to feed a hungry world are just the beginning — scientists hope CRISPR could also help shed the stigma surrounding genetically modified organisms (GMOs). In 2016, agriculture technology company DuPont Pioneer announced a new variety of CRISPR-edited corn that, because of how researchers altered its genes, is technically not a GMO.

The distinction between GMOs and gene-edited crops is fairly simple. Traditional GMOs are made by inserting foreign DNA sequences into a crop's genome, transmitting traits or properties to future organisms. Gene editing is more precise than that: it makes precise alterations to genes in specific locations of the native genome, often knocking out certain genes or changing their location, all without introducing foreign DNA.

4. CRISPR Could Eradicate The Planet's Most Dangerous Pest

Gene-editing techniques like CRISPR could directly combat infectious diseases, but some researchers have decided to slow the spread of disease by eliminating its means of transmission. Scientists at the University of California Riverside developed a kind of mosquito that is uniquely susceptible to changes made with CRISPR, giving scientists unprecedented control over the traits that the organism passes to its offspring. The result: yellow, three-eyed, wingless mosquitoes, created by altering gene responsible for eye, wing, and cuticle development.

By disrupting target genes in multiple locations of the mosquito's genes, the team is testing a "gene drive" system to spread these inhibiting properties. Gene drives are a way to essentially ensure that a genetic trait would be inherited. By impairing the mosquito's flight and vision, the Riverside team is hoping to greatly reduce its ability to spread dangerous infectious diseases among humans, such as dengue and yellow fever.

Other researchers are getting rid of mosquito populations by interfering with how they reproduce. At the Imperial College London in 2016, a team of researchers used CRISPR to target female reproduction of the type of mosquito that carries malaria through a gene drive system that influenced female-sterility traits into being more likely to be inherited.

But interfering with mosquito populations could have unforeseen consequences. Eliminating a species, even one that doesn't appear to have much ecological value, could upset the careful balance of ecosystems. That could have disastrous consequences, such as disrupting the food web or increasing the risk that diseases like malaria could be spread by different species entirely.

CRISPR TOMORROW

Current scientific advancements show that CRISPR is not only an extremely versatile technology, it's proving to be precise and increasingly safe to use. But a lot of progress still has to be made; we are only beginning to see the full potential of genome-editing tools like CRISPR-Cas9.

Technological and ethical hurdles still stand between us and a future in which we feed the planet with engineered food, eliminate genetic disorders, or bring extinct animal species back to life. But we are well on our way.

OUR NEXT AD-VENTURE

MEGASTRUCTURES OF SPACE

Humankind is energy hungry. As our civilization has industrialized over the last couple centuries, global energy consumption has spiked more than twentyfold with no end in sight. When demand outstrips what we can reap from Earth and its vicinity, what would our power-craving descendants do?

A bold solution: the Dyson Sphere. This megastructure—usually conceived of as a gigantic shell enclosing the sun, lined with mirrors or solar panels—is designed to collect every iota of a star's energetic output. In the case of our sun, that colossal figure is 400 septillion watts per second, which is on the order of a trillion times our current worldwide energy usage. What's more, the interior of the Dyson Sphere could, in theory, provide far more habitable real estate than a measly planet.

Physicist Freeman Dyson speculated that a technologically advanced race, reaching the limit of its civilization's expansion because of dwindling matter and energy supplies, would seek to exploit their sun for all it is worth.

"One should expect that, within a few thousand years of its entering the stage of industrial development, any intelligent species should be found occupying an artificial biosphere which completely surrounds its parent star," Dyson wrote in the 1960 Science paper that led to his becoming the namesake of this megastructure.

A DUBIOUS SPHERE

From an engineering perspective, a Dyson Sphere sounds pretty wild. And it is: As an immense, hollow ball, the structure is impossible. "An actual sphere around the sun is completely impractical," says "Stuart Armstrong, a research fellow at Oxford University's Future of Humanity Institute who has studied megastructure concepts.

Armstrong says the tensile strength needed to prevent the Sphere from tearing itself apart vastly exceeds that of any known material. Another problem: The Sphere would not gravitationally bind to its star in a stable fashion. This is perhaps counterintuitive; you might think that a perfect sphere around a star would be stable. But if any part of the sphere were nudged closer to the star—say, by a meteor strike—then that part would be pulled preferentially toward the star, creating instability.

That's too bad. If it could be stabilized, a Dyson Sphere built at 93 million miles from the sun, the same distance as the Earth, would contain about 600 million times the surface area of our planet in its interior. However, comparatively little of the surface would be habitable on account of a lack of gravity. By spinning the whole sphere, we create gravity in the form of centrifugal force along an equatorial band. But this rotation would wrack the megastructure with yet more destructive stress.

If the Dyson Sphere were possible, its residents would be treated to an awesome vista. The "sides" of the inner Sphere would seem to contain the observer within a bowl-like tunnel, with the sun, constantly overhead, appearing as a light at the tunnel's "end." Astonishingly, along those sides, an object the size of the Earth would look miniscule.

FLOCKING MIRRORS

Okay, so the fanciful Dyson Sphere appears to defy the laws of physics. A related concept—the Dyson Swarm—is more promising. "The Swarm is the more realistic model," Armstrong says.

A Dyson Swarm consists of thousands of relatively small mirrors or solar panels in an array of orbits around the sun. Like a dense cloud of bees buzzing around a hive, a Dyson Swarm largely shrouds the sun from external view, capturing most of the available solar energy.

A robot-driven manufacturing process could build up a Dyson Swarm in as little as several decades. His plan relies on exponential returns from a virtuous cycle beginning with robots mining material from Mercury. The material is rocketed into orbit (not too tough, given Mercury's weak gravity), then fabricated into an energy-collecting Dyson Swarm unit.

TRANSMOGRIFIED PLANETS

About half of Mercury's mass—2 sextillion pounds or so—would be usable in the form of the elements oxygen and iron, Armstrong reckons. These elements could be combined to form an iron oxide called hematite, which we humans have used to make mirrors since antiquity. The mirrors could reflect sunlight to power a generator akin to a solar thermal energy plant but adapted for operation in space.

After 40 years of getting worked over, Mercury would be kaput. The small planet would have been converted into a horde of mining and manufacturing robots, powered by fleets of Dyson Swarm solar collectors. Making a full Dyson Swarm that would catch nearly all of the sun's rays, though, would require dismantling perhaps the entire inner solar system—Mercury, Venus, Earth, and Mars. But once engineers have reached this advanced stage, Armstrong says, this prospect wouldn't seem so daunting. Strip-mining Venus would take merely a year given all the available energy and robotics following Mercury's demise.

In devising this Dyson Swarm game plan, Armstrong assumed—conservatively, he thinks—only a one-tenth efficiency for rocketing material off Mercury. The other 90 percent of available energy would go toward mining and processing ore. He assumed further that the mirror and associated generator would reap just a third of the available solar energy, less than some of today's solar concentrator efficiencies.

EVERYTHING YOU NEED TO KNOW ABOUT HYPOTHETICAL SUN MEGASTRUCTURE, THE DYSON SPHERE

If feasible, the Dyson Sphere has the potential to harness the power of the sun and send it back to earth to power all of civilization.

Even more so, the star is slowly fading over time. Now, this was strange even for the realm of the universe. Though some argue that it could be dust around the planet causing the planet to dim so sporadically, most ruled out this possibility. So, what is the alternative explanation? Aliens.

Well, not quite. But also maybe. A viable theory for the cause of this star's odd behavior, scientists believe that an alien megastructure could be the cause, a Dyson sphere. In short, a Dyson sphere would be a highly advanced piece of technology able to harness the power of the sun.

Though it sounds like science fiction, the idea of creating a Dyson sphere is very much possible and could become a reality once technology appropriately catches up with this ambitious idea. Even more so the creation of a Dyson Sphere may be the key to space travel beyond our own stars and the future colonization of humanity.

So, if you heard people talking about a Dyson sphere or potentially want to start building your own sometime in the future, here is everything we need to know about the Dyson sphere and its potential to completely change a civilization.

If this were to occur, even with just a fraction of the Sun's energy civilization would not need to heavily rely on fossil fuels or other current sources of energy.

Humanity's need for energy will increase as it evolves and makes new advances in technology. Eventually, the world will have to come up with an effective way to deal with this growing energy demand. And, what better energy source is than the sun?

The sun is going to be around for a few more billion years, again by just tapping into a fraction of this energy with the help of a Dyson sphere humanity would be able to power itself for generations.

BUILDING A DYSON SPHERE

Basically, the idea of a Dyson sphere centers around building a structure large enough to encompass the sun. Though there are many proposed ways to build a Dyson sphere there are four main lines of thinking that we should know about a Dyson Ring, a Dyson Swarm, a Dyson Bubble, and a Dyson Shell.

Dyson Ring

Think of the Dyson Ring as simply just that, a massive ring around the sun. The large energy gathering rings would be comprised of a huge number of co-planar solar sails installed around a star.

Each little sail would occupy the same plane as the Earth making the distance between them and the Sun about 1 Astronomical Unit. After the energy is collected by the sails, the energy would be transferred wirelessly back to earth.

Dyson Swarm

The Dyson Swarm takes the idea and amplifies it, very literally creating a swarm of Dyson Rings. The sun would be surrounded by countless rings full of solar sails ready to collect energy and transfer it back to earth.

However, the orbital mechanics of the task would make the creation of a Dyson Swarm very tricky to make a reality.

Dyson Bubble

The Dyson Bubble idea takes the previous ideas and builds on them creating a system of intelligent statites that would completely surround the star. Technology has nowhere near reached this level but it another great alternative way of creating a Dyson structure.

Dyson Shell

Probably one of the most commonly used lines of thinking when discussing, a Dyson structure, this megastructure forms completely around the sun with an energy-absorbing material enveloping the Sun.

However, the gravitational interaction between the star and device would be problematic, potentially causing the star to destroy the structure unless researchers create a propulsion system powerful enough to combat this issue.

LIVING IN THE DYSON SWARM ERA

If we're going to destroy the Earth to build the swarm, then obviously we'll need some habitat units amidst the Swarm. These could come in the form of large, rotating space colonies, like O'Neill Cylinders, placed at a nice, temperate, average Earth–Sun distances, and in safe zones where Swarm solar collecting units would not swoop through. The habitats could be configured to receive energy via lasers from the vast Swarm network.

Then again, creating Earthlike oases amidst the Swarm as replacements for our departed planet might not be the true motivation of a Dyson Swarm society. A commonly suggested reason why humankind might one day desire all the Sun's radiated energy is to power incredibly sophisticated computers. Maybe those computers would, in fact, be us—in the form of post-biological consciousness with no need for air, water, or food.

When thinking of how future people may view [building a megastructure], we tend to get caught up with specific images of various habitats, with natural grass and other things.

Then again, creating Earthlike oases amidst the Swarm as replacements for our departed planet might not be the true motivation of a Dyson Swarm society. A commonly suggested reason why humankind might one day desire all the Sun's radiated energy is to power incredibly sophisticated computers. Maybe those computers would, in fact, be us—in the form of post-biological consciousness with no need for air, water, or food.

"When thinking of how future people may view [building a megastructure], we tend to get caught up with specific images of various habitats, with natural grass and other things.

HUMANITY FUTURE?

Since the beginning of civilization, humanity has always strived to advance to the next level. From the development of irrigation systems to the creation of artificial intelligence, we've always been at work trying to make our lives as efficient and our tools as advanced as possible. Yet we may very well come to a point where the energy available on Earth (whether derived from fossil fuels or alternative sources) is simply not enough to meet human needs. This is where the idea of the Dyson swarm comes into play.

The Dyson swarm is a derivative of the Dyson sphere, a conceptual megastructure first theorized by Freeman Dyson. The idea of the Dyson sphere involves the large-scale collection of solar energy for human use through the construction of a massive megastructure around the Sun, built with energy-absorbing materials. The Dyson sphere, however, would not be the optimal way to capture a star's energy due to the exorbitantly high cost of creating a mostly solid sphere around the Sun—a project most likely too expensive for the resources of all nations on Earth, let alone any one country. The Dyson swarm is more plausible, as it would essentially involve individual solar panels surrounding the Sun and absorbing its light energy into solar collectors for practical use. Furthermore, this would minimize the amount of materials needed and provide a more stable structure than a project like the Dyson sphere.

The primary benefit of building the Dyson swarm is that it would provide an enormous amount of energy compared to what we as a species produce annually. According to Ibrahim Semiz and Salim Ogur, who work at the Department of Physics at a Turkish university, "[The Dyson swarm] would receive all the power of the Sun, 3.8×1026 W, in contrast to the power intercepted by Earth, 1.7×1017 W." Fundamentally, the Dyson swarm would provide us with a critical source of energy that would neither harm our environment nor run out for the foreseeable future.

A paper published by Stuart Armstrong and Anders Sandberg goes into the specifics on how to accomplish building the Dyson swarm. Their idea is to acquire the materials from the metal-rich planet Mercury to produce the solar panels and collectors by building a planetary base there. Then, an automated process would send the finished parts to orbit the Sun to be assembled later in space.

The energy needed to start the process could continuously be supplemented by using the Sun's energy at Mercury through these solar panels. As a result, the creation of the solar panels would exponentially increase, as each one would power the automation process to create another, then the two would power two more, and within a mere decade, we would be able to encircle our entire star with solar panels. If we are able to construct the Dyson swarm, it would launch us to a Type II civilization: a society that can control a star's energy. This is part of the Kardashev scale, which was created by Nikolai Kardashev to measure a civilization's level of scientific advancement in terms of energy. As of right now, we are at Type I civilization, and the Dyson swarm brings the potential of going into Type II. With the Dyson swarm, we could easily power the energy needed to travel to other planets, and the time it takes for new technological developments and ideas to be made would lessen with an essentially infinite power source.

There are several drawbacks and obstacles to the creation of the Dyson swarm: the possible disagreements and conflicts between countries due to political issues that would distract us from construction and the amount of technical innovation from automation needed to complete the process. The project may well take centuries to complete, which brings the possibility that the project may never be started. However, if in the future we were to work together as human beings striving to advance our species, then the Dyson swarm would be an efficient way of jump-starting it. The potential benefits of building the Dyson swarm would pave the way for expansion toward other planets in the solar system and allow for more technological innovation, turning what is now science fiction into reality.

THE DILEMMA OF A DYSON

The universe is a cold, heartless place. Once we've consumed all of our Earth-based energy resources, we'll be in dire need of a way to power our furnaces and refrigerators. Our sun is a like a humongous power plant, warm and life giving. It's our best shot at perpetuating our species and evolving into more capable creatures.

At present, though, a Dyson sphere of any kind is simply beyond our means. If we chose to mine Mercury, for example, we'd need robot technology that just doesn't exist at present. Those robots would need to operate flawlessly far from their human commanders, working for decades to fashion raw materials into energy collector technology. That means extracting the valuable metals from rock and then somehow building sophisticated electronics, all without on-site human help.

There's also the challenge of getting collected power back to Earth so that it can power your television. A really long extension cord probably won't cut it. People have instead suggested using laser beams or microwaves for this purpose. But lasers lose their efficiency after traveling less than a mile. Microwaves work at much longer distances (nearly 100 miles, or 161 kilometers), but nowhere near far enough for the purposes of a Dyson sphere.

Although powering our planet this way isn't a possibility at present, the concept of Dyson spheres may very well help us find extraterrestrials that have moved past the Type I stage. In 1960, Dyson figured that if a civilization did indeed manage to channel a star's electromagnetic energy, there would be a lot of leftover heat pushed outwards as a byproduct.

Detecting that outgoing infrared radiation may be the key to detecting other intelligent life forms on the other side of the universe, which researchers are currently investigating. They've already found areas with a lot of the heat of a star but without the light, leading some to think that aliens may be trapping much of the energy.

What all of this means is that we're simply stuck in the Type I civilization category for the moment. As the centuries pass, though, our technologies may advance exponentially. And if they do, we may find that we're able to turn our sun into a power source that may transform our entire race, making us more technologically proficient and space-worthy than we could've ever dreamed.

FINDING THE PLANETS

BEST FIT FOR OUR ACCOMODATION

HOW TO MAKE A SECOND EARTH

Terraforming is the process of making a planet more Earth-like. Does that mean we could make Mars, Venus, or some of the moons in our Solar System a viable place to live?

WHAT MAKES OTHER PLANET HABITABLE?

Discovering thousands of planets beyond our solar system counts as a "eureka" moment in human exploration. But the biggest payoff is yet to come: capturing evidence of a distant world hospitable to life.

A planet's habitability, or ability to harbor life, results from a complex network of interactions between the planet itself, the system it's a part of, and the star it orbits. The standard definition for a habitable planet is one that can sustain life for a significant period of time. As far as researchers know, this requires a planet to have liquid water. To detect this water from space, it must be on the planet's surface. The region around a star where liquid surface water can exist on a planet's surface is called the "habitable zone." However, this definition is confined to our understanding of current and past life on Earth and the environments present on other planets. As researchers learn more and discover new environments in which life can sustain itself, the requirements for life on other planets may be redefined.

Different types of planets may drive processes that help or hinder habitability in different ways. For example, planets orbiting low-mass stars in the habitable zone may be tidally locked, with only one hemisphere facing the star at all times. Some planets may be limited to only periodic or local habitable regions on the surface if, e.g., they experience periodic global glaciations or are mostly desiccated. In order to understand the full range of planetary environments that could support life and generate detectable biosignatures, we require more detailed and complete models of diverse planetary conditions. In particular, understanding the processes that can maintain or lead to the loss of habitability on a planet requires the use of multiple coupled models that can examine these processes in detail, especially at the boundaries where these processes intersect each other

Even if we assume any habitable planet must be Earth-like (and it may not be), chances are we're not alone. Astrobiologists estimate that the Milky Way has 500 habitable planets, which fit the following criteria:

- They're a comfortable distance away from a star similar to our sun. That is, they're far enough away to be out of the heavy heat and radiation zone, but not so far that they're extremely cold. This just-right distance is called the "habitable zone."
- They're made of rock. Jupiter, Saturn, and Uranus in our solar system are made of gases, so we don't expect life to be able to survive there.
- They're big enough to have a molten core. Earth's core gives us a source of geothermal energy, it allows cycling of raw materials, and it sets up a magnetic field around the planet that protects us from

radiation. Mars probably had a hot liquid core at one time, but because it's a smaller planet its heat dissipated more quickly.
- They are good candidates for having a protective atmosphere. The atmosphere holds carbon dioxide and other gases that keep the planet warm and protect its surface from radiation.

WHAT ABOUT INTELLIGENT LIFE?

Even today, most Earth life is microbial. And microbes have been here for at least 3.5 billion years—a very long time when you consider that the first rocks formed about 3.8 billion years ago. Given the number of potentially habitable planets within our home galaxy, chances are good that microbial life exists elsewhere.

Whether more-complex life forms like plants and insects exist elsewhere is a bigger question. For 2.9 billion years or so, all life on Earth was microscopic. Multicellular, macroscopic algae and animals came onto the scene about 600 million years ago. That's about 17% of the total time that life has existed on Earth.

It's very hard to say how likely we are to find other intelligent life forms. The oldest fossils of our species, Homo sapiens, are less than 200,000 years old. That's a geological microsecond, just 0.000057% of the time microbes have been around. If, like on Earth, intelligent life takes much longer to evolve than microbial life, it might be rare. But some galaxies and solar systems are much older than ours, so maybe other intelligent life has formed; however, it may be long dead. Finding living intelligent beings is a matter of both space and time.

HOW WILL CRISPR IMPROVE?

These developments in the CRISPR technique indicate how the technology is set to improve and develop in the future. However, research is far from over.

Platt believes that CRISPR processes are still in their infancy, as the current tools are effective at cutting DNA but can result in random repair. He believes that the future of genome editing is going to require new tools to enable more precise changes to the genome.

Eliminating random output would ensure success of the technology for therapeutic effect. Making precise changes is therefore the direction that CRISPR would evolve to, allowing more complex challenges to be tackled.

Gersbach remarks that his team's study will likely stimulate more research into Class 1 systems, which could lead to numerous applications and provide more biological insights into its potential therapeutic use.

Although there is more work to be done with regard to Class 1 CRISPR systems, its unique attributes make it worth investigating, he says.

Xu also comments that CRISPR is a weng field compared with other technologies. He highlights the many areas of CRISPR developments: better editors; larger animal or in vitro models; and more precise analytical methods to detect gene editing.

He believes that CRISPR holds tremendous potential to treat disease, which is "absolutely ground-breaking." If specific, targeted genes in the body can be controlled, then almost every condition could potentially be treated.

MAKING HUMANS A MULTI-PLANETARY SPECIES

Achieving sustainable human presence on alien planetary bodies will expand our understanding of the cosmos, our capacity to investigate fundamental questions, such as the potential for life beyond our home planet, and will enable continued growth of the global economy. Space agencies such as NASA (National Aeronautics and Space Administration) and ESA (European Space Agency) as well as companies from the private sector like SpaceX share the common interest of moving forward the human exploration of deep space and launching the first manned missions to Mars in the near future. A major factor limiting the expansion of human space exploration is the enormous logistical costs of launching and resupplying resources from Earth.

Space is the final frontier. It's expensive, dangerous, and difficult to go there. But we shouldn't forget how far we've come since our ancestors first left Africa over a hundred thousand years ago and learned to survive and thrive nearly everywhere on Earth. Humans have consistently proven that we are exceptional at pushing the boundaries of progress.

Exploring and traveling through the universe is perhaps the most impractical feat of all. Preoccupied by our everyday lives and earthly ambitions, it's easy to forget we are living in a universe 93 billion light-years across. In the estimated two trillion galaxies in our observable universe,

NEAR FUTURE

there are more stars than grains of sand on all the beaches on Earth. Think about that.

Last year alone, we saw many key highlights and progress in space exploration. Just recently, Virgin Galactic's passenger-carrying spaceship VSS Unity has completed its seventh unpowered glider test flight. SpaceX has also announced an Interplanetary Transport System and has seen many successful test launches. Nasa's chief scientist Ellen Stofan predicts that humans will be on Mars by the 2030s.

Given these trends, it's very likely that children being born today would have the opportunity to live on a different planet. But are we preparing them for it? Are we equipping our weng minds with the skills, values and mindsets to be a spacefaring species?

A cosmic citizen is anyone who recognizes our place in the universe, the fragility of our planet, and the unimaginable potential we have as a species. At its core, the cosmic perspective is about zooming out and seeing the big picture. It involves acknowledging our place in the cosmos and stepping back and contemplating our purpose in the grand scheme of things. Becoming a cosmic citizen is a powerful awakening of the mind and a fundamental redefinition of what it means to be human. It upgrades our consciousness, our values, and the kind of ambitions that we set forward for ourselves, both as individuals and as a species.

For too long, multi-planetary beings depicted in popular science fiction, seemed to carry with them the same negative traits as the current state of humanity. Sci-fi TV shows, more often than not, tell the stories of intergalactic wars, conflict and destruction. Yet, it may well be likely that the future of humanity in the cosmos is one of prosperity, peace and transcendence.

However, we would not accomplish such a feat simply by building an interplanetary transport system or more efficient rockets — we also need to invest in developing the right set of intellectual values in our youth and among society at large.

KEY SKILLS, MINDSETS & VALUES FOR COSMIC CITIZENS

The key characteristics we need to develop in our youth in order to empower them to become cosmic citizens broadly include:

1. Moving beyond being merely a global citizen and thinking of oneself as a citizen of the universe.

2. Seeking knowledge about the cosmos.

3. Being future ambassadors for planet earth.

4. Being aware of our connection with the rest of the universe biologically, chemically, and atomically.

5. Taking a larger, cosmic perspective about life.

6. Being aware of the existential threats facing our species.

We need to remember that the "Mars Generation" will be faced with unique challenges like never before. In order to prepare them to become a multidisciplinary species we need to equip them with with a cosmic perspective, in addition to:

- The desire to contribute positively to human progress.

- The agility, adaptability and risk-taking skills to survive extreme environments and existential threats.

- The grit and resilience to stay strong in the midst of potentially life-threatening, uncomfortable and radically novel environments.

- The mindset of Intelligent Optimism, so that they remember the power of human potential in the face of grand challenges and cosmic threats.

- The collaboration skills — along with values of kindness and compassion — to avoid interstellar and intra-planetary conflicts that could pose a threat to the survival of our species.

- The creativity, imagination and problem-solving to be able to create entire worlds and societies from scratch — and to provide innovative solutions to cosmic challenges.

- The existential awareness, and the emotional intelligence, that will allow them to better handle mental/internal and social/external conflicts.

- The values and mindsets that will one day, make them the ideal ambassadors when they represent earth to other intelligent extraterrestrial beings.

- The curiosity and sense of exploration to keep pushing the final frontier no matter what the circumstances.

If we are to become a successful interplanetary, and, possibly, intergalactic species, we need to instill these mindset and skills in our youth. We need to empower them to become cosmic citizens. It's exactly why Cosmic Citizenship is a core focus area in the curricula and programs at Awecademy.

Ultimately, setting such grand and powerful ambitions for our youth and our species at large will instill an invigorating sense of purpose to become life-long learners and contribute to human purpose.

CONCLUSION

People usually have more information about the near future than the distant future. They should therefore make more confident predictions regarding the near future.

Confidence in predicting near future and distant future outcomes. This book found that participants were more confident in theory-based predictions of psychological experiments when these experiments were expected to take place in the more distant future.

Effective planning often requires predicting outcomes that are expected in the relatively distant future. For example, planning a vacation often requires predicting a long time in advance the weather conditions at the destination of the trip. Similarly, class registration requires predicting a long time in advance whether the selected courses would live up to one's expectations.

The question then is whether and how temporal distance from future outcomes affects individuals' predictions about those outcomes. Does temporal distance affect the confidence with which predictions are made? Temporal distance from future outcomes ordinarily reduces the accuracy with which those outcomes can be predicted. Are individuals sensitive to these temporal differences in accuracy?

While the future can never be predicted with absolute certainty, present understanding in various scientific fields allows for the prediction of some far-future events, if only in the broadest outline. These fields include astrophysics, which has revealed how planets and stars form, interact, and die; particle physics, which has revealed how matter behaves at the smallest scales; evolutionary biology, which predicts how life would evolve over time; and plate tectonics, which shows how continents shift over millennia.

All projections of the future of Earth, the Solar System, and the universe must account for the second law of thermodynamics, which states that entropy, or a loss of the energy available to do work, must rise over time. Stars will eventually exhaust their supply of hydrogen fuel and burn out. Close encounters between astronomical objects gravitationally fling planets from their star systems, and star systems from galaxies

Physicists expect that matter itself will eventually come under the influence of radioactive decay, as even the most stable materials break apart into subatomic particles. Current data suggest that the universe has a flat geometry (or very close to flat), and thus will not collapse in on itself after a finite time, and the infinite future allows for the occurrence of a number of massively improbable events, such as the formation of Boltzmann brains

THANK YOU